I0469585

Chemistry in the World

Chemistry in the World

Brandon Weifenbach

Copyright © 2011 by Brandon Weifenbach.

ISBN: Softcover   978-1-4628-6322-8

All rights reserved. No part of this book may be reproduced or transmitted in any form or by any means, electronic or mechanical, including photocopying, recording, or by any information storage and retrieval system, without permission in writing from the copyright owner.

This book was printed in the United States of America.

To order additional copies of this book, contact:
Xlibris Corporation
1-888-795-4274
www.Xlibris.com
Orders@Xlibris.com
78849

Contents

Chapter 1—Water Treatment

Less than 1% of the liquid water on the Earth's surface is freshwater. Freshwater is what you use daily for drinking and washing. Freshwater is cleaned, or treated, so it can be again. The scientific method is used in the water treatment process. Testing and observation show that the water needs to be cleaned. Hypotheses propose ways of cleaning water. Data are gathered and analyzed. Conclusions tell whether clean water is produced. The results are shared with others. Water is treated in three steps. The first step removes solids. Stones and sand settle out of the water. Then the water passes though screens that remove solid materials that float. The water then stands in large tanks. Some solids in the water sink and are removed. Grease and floating solids are skimmed off the surface. In the second step, bacteria are used to break down plant and animal materials. The bacteria need oxygen, so oxygen is added. After these materials break down, they clump together and are removed. The third step treats the water so it is safe to go back in to rivers, lakes, or storage reservoirs. High levels of Nitrogen and Phosphorus can harm plants and animals. To remove these substances, the water is filtered through sand or stored in large pools. It might also be placed into wetlands made for this purpose.

Notes

Notes

Chapter 2—Safety Air Bags

Airbags are important safety devices in cars and other vehicles. They help protect passengers during high-speed collisions. Did you know that air bags rely on an explosion to work properly? Sensors in the car detect a collision and send a signal to the air bag. The air bag inflates, or fills, fast enough to cushion the passenger. An air bag really should be called a nitrogen gas, not air. The nitrogen gas is produced when a compound is exploded. This compound varies, but is often sodium azide, NaN_3. It is not the only substance in the airbag. Sodium azide is part of a mixture with two other substances. When Sodium azide explodes, it produces nitrogen gas, sodium metal, and heat. The heat helps to expand the nitrogen gas and inflate the airbag faster. Sodium metal can easily react with other substances, and this properly makes it dangerous. That is why potassium nitrate, KNO_3, is also in the airbag mixture. The Sodium metal and potassium nitrate react, forming more nitrogen gas and two compounds that are less dangerous. The third substance in the mixture is Silicon Dioxide, SiO_2. It is a type of sand. High temperatures melt the Silicon Dioxide into glass. The glass traps any remaining dangerous materials. All of these chemical reactions happen in a few hundredths of a second! The reaction depends on exact quantities of each substance in the airbag mixture. Scientists and engineers continue to improve the design and content of airbags. These devices have already saved many lives.

Notes

Notes

Chapter 3—High-Sulfur Coal

Coal is a fossil fuel. It forms naturally from the remains of plants and animals deep underground. Coal is burned in power plants to produce electricity. Coal is a mixture of mostly carbon compounds. When these substances are burned, they produce Sulfur Dioxide, SO_2. SO_2 is a toxic gas with a very sharp odor. It reacts with water and oxygen in the air to form Sulfuric acid, H_2SO_4. SO_2 contributes to air pollution and acid rain. These problems could be reduced by removing the removing the sulfur from coal before burning it. However, the energy and cost to do this is great. Another solution is to burn coal that has a low sulfur content. However, a large part of the world's coal supply is high in sulfur. A more practical solution is to remove the SO_2 after burning the coal. This is done by "scrubbing" the gases produced when coal is burned. Scrubbing involves a series of chemical reaction. Calcium Carbonate, $CaCO_3$, is added to the furnace. When heated, it breaks down into Calcium Oxide, CaO, and Carbon Dioxide, CO_2. The CaO reacts with some of the SO_2 to produce Calcium Sulfite, $CaSO_3$, an ionic compound. The remaining SO_2 reacts with a mixture of CaO and water. This forms additional $CaSO_3$. The $CaSO_3$ slowly reacts with Oxygen, changing into Calcium Sulfate, $CaSO_4$. $CaSO_4$ does not dissolve well in water. It is not very acidic. Because of these properties, it can be safely burned without causing a pollution. However, $CaSO_4$ is useful as a building material. It is part of gypsum, the major component of dry wall.

Notes

Notes

Chapter 4—Desiccants

Water is a common and important compound. It is needed for life. However, there are some situations when water should not be present. For instance, some medicines break down if exposed to water for too long. Lenses and mirrors in some equipment need to be kept dry at all times. Water in the air can cause mold to grow on these glass surfaces. To keep the air dry, desiccants are used. Desiccants are often ionic compounds. They easily form hydrates. When water molecules are present, desiccants trap them. Several water molecules can combine with one formula unit of desiccant. This makes desiccants effective drying agents for the surrounding air. A common desiccant is silica gel. This is a form of Silicon Dioxide—the molecular compound in sand and glass. Silica gel is placed in packets or plastic containers. These are then placed with products that are affected by moisture, like medicines or electronics components. Silica gel packets are placed in boxes of new shoes to remove moisture. They are used around certain foods to prevent mold and spoilage. Desiccants are used on a large scale in some industries. For example, natural gas consists mostly methane (CH_4) and can combine with water vapor. Desiccants, such as Calcium Chloride or Lithium Chloride, are used near natural gas to remove water vapor. These desiccants are replacing older drying methods that wasted large amounts of Methane. Using desiccants saves fuel, reduces operation cost, and is better for the environment.

Notes

Notes

Chapter 5—Combinatorial Chemistry

Combination reactions are among the most important reactions in applied chemistry. During most if the 1900's, chemists combine molecules to create compounds called synthetics. Some synthetics are new molecules, while others are identical to molecules that occur in nature. Many pharmaceutical drugs and medicines are synthetics. Plastics are also synthetics. Plastics are long chains of molecules. Creating new synthetics is a major part of industry today. Improving ways to make known synthetics is also important. Chemical engineers look for more efficient and cheaper ways to make synthetics. Combinatorial Chemistry is a technique that is changing how compounds are produced in industry. This technique was developed in the 1980's. It was first used for producing proteins. Before this, large molecules at a time. These singe combination reactions are reliable, but not very efficient. In combinatorial chemistry, mixtures of compounds are combined with mixtures of other compound. Many combination reactions occur at once, producing many products instead of just one. The products are then carefully separated. Some have uses, while others are waste products. Combinatorial chemistry is a new technique in the pharmaceutical industry. Many drugs with similar structure can be produced at once. However, the large variety of products requires more analysis than single-product reactions. Also, better ways to separate the products are needed.

Notes

Notes

Chapter 6—Luminal

Forensic science uses scientific concepts to solve crimes. Technology is an important part of forensic science. Computers are used to see a small part of a picture or hear a certain noise in a recording. Chemistry is also an important part of forensic science. Chemical tests are used to identify evidence found at a crime scene. One substance used at many crime scenes is luminal. Luminal is a compound that contains nitrogen, hydrogen, oxygen, and carbon. Before using luminal, a forensic scientist mixes with certain amounts of hydrogen peroxide and other substances. The substances in this mixture do not react with each other unless a catalyst is present. A catalyst is a substance that speeds up a reaction. The catalyst that is needed to cause a reaction is hemoglobin. Hemoglobin is a compound found in human blood. Suppose that during a crime, a person is injured. The criminal cleans the area so that no blood is seen. Later, forensic scientists spray the luminal mixture around the crime scene. If any blood is present, even in very small amounts, the hemoglobin will cause the luminal and hydrogen peroxide to react. The product that forms will glow enough to see at night or in a dark room. The glowing products show where blood is present. A few other substances, such as Chlorine bleach, also cause the mixture to react. Trained forensic scientists can usually tell if blood caused the reaction by how quickly the glowing appears. Other chemical test can prove that the glowing was caused by blood.

Notes

Notes

Chapter 7—Gas and Fires

A fire needs fuel, oxygen, and high temperature. If anyone of these things is missing, the fire does not burn. Some fuels are gases. Hydrogen gas, for example, burns. Common fuels such as methane and propane are gases. Oxygen is a gas that combines with two other gases: water vapor and carbon dioxide. Gases are often used to extinguish, or put out, fires. These gases are non-flammable, which means they do not burn. Instead, they remove at least one of the three requirements for a fire. Some fire extinguishers contain a liquid under pressure. This liquid changes to a gas and expands as it leaves the extinguisher. This change of state cools the gas. The cooled gas lowers the temperature of the fire. The gas coming out of a fire extinguisher is also denser than air. It acts like a blanket over the fire, keeping oxygen away from the fuel. Some extinguishers spray dry chemicals or water to put out a fire. Even these extinguishers use a dense, nonflammable gas to push the chemicals or water out of the tank. Carbon dioxide is commonly used in fire extinguishers. This gas does not burn, is denser than air, and is inexpensive to make. Halon was used in fire extinguishers for many years. This gas extinguishes fires without leaving materials behind. However, halon harms the environment. It can no longer be used in fire extinguishers.

Notes

Chapter 8—MRI

Since their discovery in the later 1800's, x-rays have been used to diagnose medical conditions such as broken bones and gum disease. However, x-rays have limited use. They expose patients to harmful radiation. They cannot produce images on soft tissues such as muscle and nerves. X-ray images are often cloudy. These drawbacks have been overcome by a more recent diagnostic tool called Magnetic Resonance Imaging (MRI). MRI works because protons in a magnetic field produce radio waves. How does this happen? Water and most tissues in the human body contain hydrogen atoms. The nuclei of these hydrogen atoms are protons. A proton behaves like a magnet. The magnetic fields of protons usually cancel each other. However, when protons are placed in a strong magnetic field, they align with it. The alignment is not perfect, and the protons wobble. This wobbling produces radio waves at a specific frequency. Frequency is the number of waves produced each second. An MRI machine applies a magnetic field in a specific part of the patient's body. The protons in this part of the body wobble and emit radio waves. Then the machine applies a short pulse of radio wave at a matching frequency. This temporarily knocks the protons out of alignment. When the applied radiowaves are gone, the protons realign and produce more radio waves. The MRI machine records the frequency and intensity of these waves. The strength of the magnetic field changes, and the new data are recorded. Then the direction of the magnetic field is changed, and the whole process is repeated. An MRI scan records data along three directions. A computer assembles these sets of data into three dimensional images.

Notes

Notes

Chapter 9—Radioimmunoassay

Radio tracers have made it possible to diagnose many medical disorders. For example, Holmium-166 is used to examine liver tumors. Iron-59 is used to study the spleen. Iodine-131 has long been used to diagnose diseases of the thyroid gland. These radioisotopes have short half-lives, and small amounts are given to patients. Another tracer techniques does not involve placing radioisotopes inside the patient. Instead, it requires only a sample of blood or other fluid from the patient. This technique is called radioimmunoassay, or RIA. It was developed by Rosalyn Yalow and Solomon Berson in the late 1950's. RIA uses tiny quantities of radioactive material. This reduces the problem of tracking and disposing of large amounts of radioisotopes. RIA uses reaction that take place in the human immune system. The immune system makes antibodies, which are proteins that react with and fight antigens. Antigens are foreign molecules that activate the immune system. In RIA, radioisotopes are attached to antigens and used as "tags." A known amount of tagged antigens is separated into several samples. Varying amounts of untagged antigens are added to the samples. The tagged and untagged antigens compete to react with the antibodies. Fewer tagged antigens react with antibodies as the amount of untagged antigens increases. For each sample, the radiation from the tagged antigens is measured. This indicates the amount of untagged antigens in the sample. This information is used to find an unknown antigen concentration in a patient sample. RIA is so sensitive that antigen amount equal to a trillionth of a mole can be detected. RIA is used to diagnose allergies and detect viruses in donated bloods. RIA is also used to determine if certain hormone levels in the blood are normal.

Notes

Notes

Chapter 10—Spectroscopy and Pollution

Emission spectroscopy are instruments that can identify almost any substance by the pattern of light it emits. Every substance produces a unique pattern. The location of bands of light identifies the substance. The size of the bands tells how much of the substance is present. These instruments have many uses. They are used to identify traces of substances left at a crime scene. They are used to identify the substances that make up a consumer product. They are also used to analyze air and water samples for pollutants. By analyzing water samples taken at different locations, scientists can locate the source of pollution as well as identify the pollutant. Air samples can be analyzed in the same way. Air near a landfill can be analyzed for pollutants. Emissions produced by factories can be checked. Animal feed lots can be monitored to be sure they do not pollute air and water. Many spectroscopes require a sample to be tested in a vacuum. This can affect the results for samples that contain water. When a vacuum is used, water evaporates from aqueous samples. The evaporation changes the concentration of any solutes dissolved in the sample. Scientists and engineers have now developed emission spectroscopes that do not require a vacuum.

Notes

Notes

Chapter 11—Arsenic and Semiconductors

Arsenic is not abundant, but it is widely spread throughout the earth's crust. Because it exists in ore containing gold and silver, Arsenic was discovered by the oldest civilizations. The poisonous properties of Arsenic make it an effective pesticide. Compounds of Arsenic have been used to treat certain diseases. Today, Arsenic is important in making electronics components that contain semiconductors. Semiconductors are substances that conduct electric charges better than insulators, but not nearly as well as metals. Germanium and Silicon are semiconductors. However these elements become more effective conductors when Gallium or Arsenic is added. Because of this, the electronics and computer industries use large quantities Arsenic. Much of this Arsenic is washed away from the newly made semiconductors during cleaning. High concentrations of Arsenic in waste water from manufacturing plants can cause illness or even death. A solution is to recycle the waste water and remove the Arsenic. One way to do this is to use fine grains of iron oxide. Another way is to cause the Arsenic in the water to become part of a solid compound. Recycling Arsenic keeps it from reaching public water resources. The removed Arsenic is used again to modify semiconductors. The methods for removing Arsenic are not limited to contaminated waste water. In some areas, the drinking water has naturally high concentrations of Arsenic. Removing the Arsenic makes this water safe to drink.

Notes

Notes

Chapter 12—Petroleum and its Molecules

Petroleum is a natural resource that supplies much of the world's energy needs. Many fuels, such as gasoline and heating oil, come from petroleum. Petroleum is also the source of most plastics and synthetic fibers. What plastic items do you see? Which pieces of clothing are made of synthetic fabrics? They all came from petroleum. Even items not made from Petroleum are often produced using Petroleum as an energy source. Petroleum is a mixture of molecular compounds called hydrocarbons. As the name implies, hydrocarbons contain hydrogen and carbon. Some of the hydrocarbons in petroleum are small molecules and some are extremely large molecules. Not all compounds in petroleum are equally useful. When Petroleum is refined, large molecules that are less useful may undergo a process known as cracking. Cracking involves breaking some covalent bonds in large molecules. Refiners use this process to change large molecules into smaller, more useful molecules. Cracking is done by two different methods. Sometimes, heat and pressure are used to break the bonds. More commonly, a catalyst is added to the Petroleum to help break large molecules into smaller ones. Sometimes, Chemists join small hydrocarbons to create large molecules. Polymerization is one such process. In polymerization, small molecules from petroleum are placed under high pressure and heat. This causes the molecules to bond. The large molecules that form are used to make gasoline.

Notes

Notes

Chapter 13—Hydrogen Power

Most of energy people use comes from burning fossil fuels. Petroleum is a fossil fuel. It is refined into gasoline and other products. Burning gasoline in vehicles produces water, carbon dioxide, and carbon monoxide. It also produces nitrogen compounds from the nitrogen gas in air. Most of these compounds contribute to air pollution. To reduce this pollution, researchers are investigating the use of hydrogen as a fuel. Hydrogen burns cleanly, producing only water. A given mass of hydrogen produces about three times the energy as the same mass of gasoline. Furthermore, the earth's water provides a practically unlimited supply of hydrogen. So, why isn't hydrogen power being used? The use of hydrogen fuel has several disadvantages. One is storage. Hydrogen gas has a low density of about $0.0139/cm^3$ at room temperature and a pressure of 160 atm. Even a liquid, its density is only $0.071 \ g/cm^3$. Liquid hydrogen must be stored below 20K, which is not practical. Very large tanks and possibly very low temperatures would be required to store useful quantities. Another disadvantage is that hydrogen is highly reactive. Almost any mixture of Hydrogen gas and air is explosive. Also, prolonged exposure to Hydrogen causes many metals to become brittle. In nature, almost all hydrogen occurs in compounds. Thus, elemental hydrogen must be manufactured. One way to manufacture hydrogen is to use electricity to break down water. However, most electricity comes from burning coal and oil. If you have to burn fossil fuel to produce hydrogen fuel, what's the point? Hydrogen fuel holds great promise for the future. However, scientists and engineers must overcome its many disadvantages first.

Notes

Notes

Chapter 14—Phase-Change Materials

During a state change, energy is absorbed or released while temperature stays constant. Some products use the energy difference between a solid and a liquid to control temperature. These products contain materials known as phase-change materials. For example, phase-change materials are sometimes added to wallboard. A common material in phase-change wallboard is paraffin, a waxy substance. When a room's temperature rises above normal, the paraffin melts slightly. As the paraffin melts, it absorbs and stores energy from the room. When the room's temperature drops below normal, at night for example, the Paraffin freezes. As it freezes, it releases energy into the room. If the outdoor temperature stays below the indoor temperature, the wallboard can be heated by electricity at night, when electricity is not in high demand. This keeps the paraffin in a liquid state. The paraffin is then allowed to freeze during the day, releasing stored energy into the room. Phase-change materials have many other uses. When packing a lunch, you may include a special pack to keep the food cold. This cold pack contains a phase-change material. It is stored in the freezer until it is needed. Then, as the phase-change material melts, it absorbs energy from the lunch contents, keeping the lunch cold. Phase-change materials are also used to keep food warm. You may have also seen or used pouches designed for this purpose. These pouches are first heated in a microwave over, which melts the solid material inside. Then they release heat for hours as the phase-change material again become a solid.

Notes

Notes

Chapter 15—Artificial Blood

A main concern of hospital is whether they have enough blood for transfusions. Even if blood is available, it may not match the blood type of the patient. Blood for transfusion is rarely available in remote locations. Also, blood can only be stored for about 42 days. So a blood surplus can soon become a blood shortage. In the 1960's, researchers began to develop blood substitutes that could be used during blood shortages. The first compounds developed were perfluorocarbons. Oxygen is 100 times more soluble in perfluorocarbons than in blood plasma. This meant that perfluorocarbons could deliver oxygen to body tissues more efficiently than the hemoglobin in real blood. However, these early blood substitutes caused health problems in many patients. Some even died. Learning from early failures, scientists improved artificial blood. Better technologies and new molecules have lead to safer blood substitutes. One promising blood substitute is being tested. Another form of artificial blood is derived from real blood. This artificial blood is a powder that can be mixed into liquid when needed. Artificial blood has some advantages over stored natural blood. It has a shelf life of nearly two years. It can be given to any patient, regardless of their blood type. Infectious diseases cannot be transferred to a patient from artificial blood. However, blood substitutes have drawback. Artificial blood only transfers oxygen, as hemoglobin does in real blood. It lacks the other components of real blood. Artificial blood can only be used for short times. Also, because of the cost of research, artificial blood is expensive.

Notes

Notes

Chapter 16—More Gasoline with Catalysts

Petroleum, or crude oil, is a micture of many different compounds. These compounds vary from simple ones containing a few atoms to complex ones containing hundreds of atoms. Some of the products obtained from crude oil are gasoline, kerosene, diesel oil, and heating oil. These products contain different mixtures of compounds. Crude oil is distilled to separate it into its components. This separation works because each compound has a particular boiling points consist of the largest molecules. These compounds, called the residue, are the least useful as fuel. Rather than throw the residue away, it is processed to produce additional gasoline and other useful products. The residue from the distillation is mixed with a catalyst and heated. This process is called catalytic cracking. The heat and the catalyst cause the large molecules to break apart. The result is a mixture of smaller molecules, such as those used in gasoline. Some Carbon atoms stop the catalyst from working because they cling to it. The catalyst is restored by heating it to burn off the Carbon. The mixture produced by Cracking is then distilled as if it were crude oil. The resulting compounds are mixed with those produced by the Crude oil distillation. By using catalysts, refineries like the one shown can produce more gasoline and other products from a barrel of oil. Catalysts also allow refineries to produce these products with less waste and at a lower cost.

Notes

Notes

_____

_____

_____

_____

_____

_____

_____

_____

_____

_____

_____

_____

_____

_____

_____

_____

_____

_____

_____

_____

_____

_____

_____

Chapter 17—Acid Precipitation

All rain is slightly acidic because carbon dioxide and water from carbonic acid in rainwater. This acid is weak, and the solution is dilute. Normal rain has a pH of about 5.6. Substances in the environment can neutralize this amount of acid. This precipitation is not acidic enough to harm the environment. However, some rain is acidic enough to damage organisms and other parts of the environment. Acid precipitation is the name given to rain, snow, or sleet that is more acidic than normal rain. Acid precipitation forms when certain pollutants are oxides of nitrogen and sulfur that are produced when fossil fuels burn. When these oxides dissolve in rainwater, they form sulfuric acid and nitric acid. Acid precipitation can make water in lakes and streams acidic, killing organisms that live in the water. It can also damage plant life and react with many types of stone. The best solution is to prevent acid precipitation from occurring. Areas where air pollutants are strictly controlled have less acid precipitation. Materials that neutralize the acids may also be used. For example, lime, CaO, neutralizes acids. Sometimes lime is added to soil and water to reduce the effects of acid rain.

Notes

Notes

Chapter 18—Biomediation

Many scientists and engineers look for ways to eliminate harmful substances from the environment. These substances range from stored toxic waste to oil spills at sea. One way to solve these problems is bioremediation. Bioremediation uses redox reactions that small organisms perform. These redox reactions change the chemical makeup of the pollutant, making it less reactive or breaking it down into safer compounds. By speeding up these usually slow reactions, the pollutant is destroyed quickly. Aerobic microorganisms are used to treat oil spills. They use oxygen to decompose the hydrocarbons in the oil. Added nutrients keep the microorganisms alive. Some pesticides and solvents contain carbon-halogen compounds. These compounds create health risks in underground water supplies. They are difficult and costly to remove directly. However, certain microorganisms obtain energy by decomposing these pollutants. When these microorganisms and hydrogen are added to a water supply, the pollutants are safely removed. Bioremediation is also used to remove toxic hydrocarbons from soil. For example, some types of fungi produce enzymes. These enzymes break down certain hydrocarbons. Some plants are able to decompose hydrocarbons in soil. Certain microorganisms that break down hydrocarbons are kept alive with nutrient-rich organic matter. Bioremediation can be supplied to many contamination problems. It can be used where the contamination is difficult to reach. In general, bioremediation techniques are safe and inexpensive. However, they cannot solve all contamination problems. In situations requiring the removal of heavy metals, for example, live organisms work poorly.

Notes

Notes

Chapter 19—Octane Ratings of Gasoline

When you pull up to a gasoline pump, you have choice of fuels with different octane ratings. What do these values mean? Gasoline is a mixture of several midsized hydrocarbons. An octane rating is a measure of how well the mixture keeps a car engine from "knocking". Knocking is caused by uneven burning of fuel in the engine. On the octane scale, a 0 value represents the large amount of knocking that occurs when unbranched heptane is burned. An Octane rating of 100 represents the very small amount of knocking that occurs when 2, 2, 4-trimethylpentane is burned. This hydrocarbon is a chain of 5 carbon atoms with three 1-carbon branches. Two of these branches are on the second carbon atom in the chain. The other branch is on the fourth carbon atom in the chain. Other octane rating are based on how well a gasoline mixture burns compared to these standards. For example, an Octane rating of 92 means that the fuel burns the way a mixture of 8 parts of heptane and 92 parts of 2, 2, 4-trimethylpentane would burn. In general, branched hyrdrocarbons have a higher octane rating than hyrdrocarbons with few branches. Aromatic compounds have even higher octane ratings. When gasoline is prepared at a refinery straight-chain hydrocarbons can be changed to branched isomers by heating the vapors with aluminum chloride. Another refining process uses catalysts to change alkanes to aromatic compounds. Usually, fuel with a higher octane rating provides better mileage—more miles per gallon. However, you should always check the owner's manual for the vehicle. Some engines are designed to operate most effectively on fuels with a low octane rating.

Notes

Notes

www.ingramcontent.com/pod-product-compliance
Lightning Source LLC
Chambersburg PA
CBHW021914170526
45157CB00005B/2070